50 Questions An[s]
More Experienced Roller Flier

By Graham Dexter

Text copyright © 2013 Graham Dexter

All rights reserved

Published by LaLas

1st Edition published December 2013

ISBN 978-1-910148-03-7

Acknowledgements:

It is with great pleasure that I have produced this new book on Birmingham Rollers, and I believe it will be a great benefit to both the experienced roller fancier and the novice starter. However, I would not have been in a position to write such a volume with the help and support of many people. In the hobby, and over many years, I have had help, guidance, direction, criticism and been laughed at by many. Most of these fanciers, whether intentionally or not, have helped form my views, spurred on my enthusiasm, sharpened my insight and generally made me a better Birmingham Roller fancier. I thank you all! My special thanks go to those fanciers, some now passed, those that consistently and persistently did and do everything in their power to produce, train, and fly the very best Birmingham Rollers that they could. It may be unfair to name individuals particularly, but I feel deeply indebted to Dave Moseley and Bill O'Callaghan particularly, for without their help I may not now still be in the fancy. When my ebb was low, they are the two people who helped me most and I am truly grateful. This book is dedicated to them both.

I would like to acknowledge the many fanciers from the USA, Canada and South Africa for the interesting conversations about rollers that further informed my views. It was their interest in learning, and through various e-mails which sought information that inspired the production of this volume format. It is also necessary to publicly pay homage to the greats such as Bill Barratt, Bob Brown, Ernie Stratford, Ken White, Bill Pensom and of course Ollie Harris.....they were in one way or another my mentors and are the reason for all our successes.

I am indebted to my good friends in the hobby like George Mason (who I have quoted a few times in this volume), Les Bezance, Deano Forster, Morris Hole, Gordon Forbes, Johnny Conradie, and my late best buddy in the hobby George Kitson. Without these people my life in the hobby would have been much much the poorer. More recently I have had help and support from Jodie Rixon and the members of the Steel City Roller Club, the Yorkshire Performing Roller Society and the All England Roller Club. My personal thanks go to John Hall and Steve Taylor for continuing to keep the National Birmingham Roller Association going from strength to strength.

I would also like to thank those people who do things for the hobby in the background for very little recognition or thanks. For example, my proof reader and daughter Kez Gibbons, and all the administrators, treasurers, ring secretaries, trophy secretaries, fly managers and club members who make this fantastic hobby the great one it is!

Wishing you all the success that you earn and the pleasure you will derive from it,

Yours in sport,

Graham Dexter

Table of Contents

General

1. How can I do better in competitions?
2. So what does it take to be a good Birmingham Roller judge?
3. How can I detect faults in my birds earlier?
4. How do I know how good my birds are?
5. What are the worst mistakes a roller fancier can make?

Flying for competition and flying for pleasure

Competition flying

6. What can I do about poor kitting?
7. I have a lot of birds rolling from the back of the team - can I change this?
8. My birds roll sloppy one day then they are okay the next - what can I do?
9. My birds roll down when trying to land, is there any cure for this?
10. I have really deep rollers, what can I do to help them in competitions?
11. My birds look 'wingy' when they roll, is there a cure for this?
12. My birds roll great for about 5 minutes then just go flat - what can I do?

13. My birds can all roll, but they don't seem to roll together - what can I do?
14. What can I do about the deep rollers? And what is the best depth to do well in competitions?
15. What can I do when the kit goes flat?
16. How can I stop my kit from flying too long?
17. How can I avoid flyaways?
18. What can I feed for slower flying?
19. How do I stop the kit splitting?
20. How can I stop the kit pinning out?
21. What can I feed for longer flying?
22. How do I balance a kit?
23. What can I feed for lower flying?
24. What are the secrets of getting the kit ready for a competition fly?
25. What is the judge looking for?
26. How many birds do I need to do well in competition?
27. I have some great rollers but a few leave the kit and fly above them, is there any cure for this?

Flying for pleasure

28. How can I get my locked up rollers flying fit again?
29. How can I avoid roll downs?
30. How many pairs should I breed from if I'm not interested in competition?
31. What's the best way to learn more about the Birmingham Roller?

32. What is the best feeding regime for youngsters, and does it differ in old birds?
33. My rollers are great as youngsters but turn to water as yearlings - is there a cure?
34. How often should I fly my kits?
35. What effects will the weather have on flying my birds?
36. How can I help the birds in the moult?
37. How long can I expect a kit to perform at their peak?

Breeding for consistent high quality rollers

38. What's the difference between 'in breeding', 'line breeding' and 'out crossing'?
39. What are the advantages and disadvantages of the Bull cock method?
40. Inbreeding: is it really necessary? Is outcrossing better than line breeding?
41. How do I know the stock is the best I can get? How do I find the best stock?
42. If you were starting again how would you go about breeding a family of rollers?
43. What are the qualities necessary in a flying bird before stocking it?
44. Should I go for quality or frequency?
45. How can I test out my breeding without breeding lots of birds?
46. What are the faults in flying birds that will show up in the breeding loft?

47. How long should I persevere with the stock I have?
48. What's the best method of selection?
49. What are the qualities in a roller that makes it a good breeder?
50. Is there a good proven breeding programme I can use?

Less experienced fanciers keep asking me questions, what should I do?

What about the missus?

General

1. How can I do better in competitions?

There are several useful tips later in this volume on how to acquire better stock, how to feed differently to avoid some of the common faults that occur on fly day, and a regime for feeding and balancing a kit. Read on to see these. In the meantime the answer to the question above is a bit more philosophical. It is important that the Birmingham Roller flyer understands the bird, understands their own motivation to compete, in order to generally get better at flying Birmingham Rollers. Firstly get to know your birds - most fanciers don't observe their birds very well, and to do well in competition you have to know the kit you intend to fly very well. If you can't close your eyes in bed at night before you go to sleep and recall all 20 of the birds you intend to fly - then you haven't even passed the first step of knowing your birds. As you go over the colour and gender of each bird in your kit, see if you can also recall just how and when they roll in the team. Are they short and frequent or deeper and less frequent? Do they roll from the front or rear of the kit? Do they roll with the team every time the kit does or do they roll just after, just before or sometimes not at all? Are they the first to drop, or do they drop with the team, or after the team has alighted? What are they like with the team in the kitbox? Are they quiet, flying from perch to perch, dominant in the kitbox, or being passive? Good tip here: if the cocks in the kitbox are boisterous and noisy there is a good chance that your kit is not going to do very well in competition that day.

When you are completely familiar with each bird in the competition kit, you are now in a position to give them the individual attention they may need. You can see the 20 birds as a team and treat them as such, or you can see the team as 20 individuals each of which may need slightly different feeding, weight gain or weight reduction, flying more or less often. You choose, but if you want to do better in the competitions you will need to consider all the things I have written here at least.

2. So what does it take to be a good Birmingham Roller judge?

Knowing the rules is perhaps the start of being a good judge. Good eyesight, and the ability to concentrate for long periods on a moving target of 20 Birmingham Rollers. Knowing what the standard is for style and performance of a perfect roll, and being able to see this in the 20 individuals as well as being able to count the birds in the breaks. There are a few things which I have spotted over the years which can prevent someone being a good judge. These are worth highlighting:

- Judging the person not the Birmingham Rollers. Personal opinion of the competitor, either for good or bad, should not enter into the scoring of the kit.

- The rules say that more than 5 birds rolling 'spontaneously' together should score points. I notice that some judges seem to give up judging a break if 2 or 3 birds start the break and ignore the 5 or more that follow spontaneously. They justify this by saying it was a 'follow through' - this is just laziness, any time 5 or more birds roll in good style spontaneously should score a break.

- 'I can't see the finish of the roll so I can't score it' - is another weak and lazy way of judging. The finish of the roll needs to be taken into account only for the attributing of quality points (or in the American system the depth and quality multipliers). If the roll has been considered good enough to score, the finish of the roll is somewhat irrelevant for break scoring.

- Some judges in the English and the American system seem to award quality points (or quality and depth multipliers) in a rather haphazard way. I often see quality points being given to too many kits without much thought. Allocating 30/200 or 1.2 for quality and 1.2 for depth for every kit that is mediocre seems to me to be a waste of useful points. In the English system, a score of 30 quality points means that there were 3 birds in the team of 20 which were of a high quality. It should have absolutely no bearing on whether the kit was able to score a break, or what the `general' quality of teamwork, kitting or style/pattern of flight was. Too many judges seem to award points to avoid criticism from the competitor rather than take it seriously and observe the individual as well as the team breaks and award points accordingly.

3. How can I detect faults in my birds earlier?

I suppose the most obvious answer to this question is careful observation! However, each bird will need to be assessed properly and that will take time. Poor kitting, and birds spending too much time away from the kit are easy to spot, however even this can be diagnosed too quickly. Some youngsters can stabilise after a few weeks and become good performers. Birds not kitting can be caused by a heavy moult, advancing years (getting craftier to avoid having to roll), or soreness due to an injury. I had a bird flying in my kit last year that was 10 years old, and was both a frequent and proficient performer – this particular bird was 4 years old before she began to roll. She was probably the luckiest of all birds, as I would not usually keep a bird more than 18 months if it hadn't started to roll. However, this bird was selected for a feeder, then didn't lay, then for some reason I tried her with eggs out of sequence which she took to and reared. Following this I continued to use her very often as she could be relied on to sit anything at any time. I would fly her out each year, and each year nothing: not even a flip, until on her fourth year she suddenly began to roll, and roll perfectly and frequently. It was only this year that she started to be more difficult to get fit, started to drop early, and began to lose favour.

So the moral of this story is, that we can be too quick to fault our birds. However, generally birds that develop faults in the first 5 months of their life and don't straighten up within a month or so, are not going to change. Unfortunately for the impatient fancier, some birds will not develop faults until they are yearlings, and some really excellent rollers may not begin to roll until they are in their second year of life. Many high quality rollers do not get the chance to show their merits, and some rollers are already in the stock pen before their owners realise their potential to perpetuate their hidden faults. It is my opinion that the following faults do not get better and therefore should not be considered for breeding:

- Poor kitting lasting for more than 1 month without specific cause.
- Loose rolling – when the bird is in good condition.
- Rolling down when trying to land (once might be acceptable, twice is marginal, persistence is proof of poor quality.
- Deep rolling and not returning to the kit speedily.
- Flying above the kit persistently (even if only 10 feet above).
- Changing direction, speed, or compactness in the duration of the roll.
- Turning away from the team at the end of the roll.
- Winginess in the roll – the birds' inability to hold its wings in the correct position during the roll.
- Laziness – i.e. persistent early dropping – check for physical defects, poor feather, respiratory illness, overweight, undernourishment first.
- Struggling to exit the roll or tail riding at the end of the roll.
- Roll downs.

4. How do I know how good my birds are?

As these 50 questions are for the more experienced fancier, this should not be a question asked very often – however! It is a shame that some fanciers who have kept rollers for some time, and certainly know the basics of keeping healthy rollers, breeding rollers, and can construct a loft suitable, may still not know just how good their birds are. The best way of discovering just how good your birds are is to visit other lofts and allow visitors to yours. By comparing what you put in the sky with others you may discover yours are superior or inferior in quality. Sadly, in my experience, many fanciers become loft blind and begin to believe that their birds are the best. Some, even experienced, fanciers have never developed their ability to discern what a good quality roller should do in the air. Observing quality rollers is the only way fanciers can compare their own stock to the very best. Of course, if you have no access to quality rollers, this becomes more difficult. As well as the following criteria there are now many examples of quality rollers on the internet which now most fanciers have access to. Pensom wrote the standard for the Birmingham Roller, which has stood the test of time, it reads like this:

> **The true Birmingham Roller should turn over backwards with inconceivable rapidity for a considerable distance like a spinning ball.**

My criteria for the perfect and proper roller is as follows:

- High velocity – the revolutions of the bird must be uncountable.
- Straight – the bird's line of descent should not deviate from plumb line straight. Clearly this may be more difficult in strong winds, but in principle the bird should roll in a straight line.
- The roller should shape in a compact round shape, not like a catherine wheel or jaggedly (caused by the wings held in the wrong position).
- The speed of the roll should remain constant throughout the roll.
- The bird should start snappily.
- The bird should stop snappily.
- Stability and control over the roll. Birds that look like they have to roll, and have little control over where or when they roll are deficient in control.
- A good frequency rate. The standard is 1 roll per minute, this being the rate when the bird is in a proper condition i.e. not in the moult, overweight, or down in condition. Recently this standard rate has been increased by some fanciers, which can only be commended providing this is not at the expense of the other factors which control quality.
- The duration of the roll (depth) has long been debated in the fancy. I favour George Mason's definition of depth of roll. He states that the roll should be: 'long enough for the spectator to evaluate the quality of the roll'. I would add that although this should be determined by the

fanciers' preference, if they do not roll for between 1½ and 2½ seconds they are probably too short to be classed as Birmingham Rollers.

- To be classed as a perfect and proper Birmingham Roller, the bird must be consistent in its performance. Once again I quote George Mason, once heard to say 'they don't roll good every time' – however, in my view they must attain a certain consistency of performance before they can be credited with quality. I would say that a Birmingham Roller of quality (when in the proper condition) should perform a quality roll as determined above at least 7 out of each 10 times it rolls. Otherwise it does not deserve the title `Birmingham Roller'.

5. What are the worst mistakes a roller fancier can make?

Fanciers' worst mistakes are listed below, please bear in mind these are my views alone, many might disagree:

- Not watching their birds enough. If they are in the air, why aren't you watching them? Why did you let them out? Of course they may need exercise when it's not possible for you to give them your full attention – however, these should be the few exceptions not the general rule.

- Staying at home, and believing your birds are the best they can be.

- Not watching other fanciers' birds – even when they are clearly not great quality, you can still learn from the errors of others.

- Being so convinced you can't improve that you stop experimenting.

- Giving away or selling too many birds which deprive you of maintaining and improving your own stock (my own particular consistent mistake which I am working on currently).

- Keeping substandard birds in order to compete in particular competitions.

- Selling or giving away birds which are substandard without a full description of their deficiencies.

- Trying to breed from too many pairs in order to boost numbers. Experiment by all means, but with care. Breeding from too many pairs usually means that a number of the youngsters won't be of great quality.

- Outcrossing for hybrid vigour. This can work in the short term, but unless done with care and expertise, second and third generations of progeny will likely be substandard.

- Haste - expecting instant results. Breeding and flying Birmingham Rollers takes time, Bob Brown once said `it takes at least 4 years from starting up with rollers to get anywhere'. I believe he was right - although it's possible to buy success through purchasing a kit from a successful fancier, it will take at least 4 years and probably a lot longer to maintain that success for long enough to breed your own family of birds, and learn all there is to know about the art of feeding, training and flying Birmingham Rollers!

Flying for competition and flying for pleasure

Competition flying

6. What can I do about poor kitting?

This is one of the greatest complaints that roller fanciers have the world over. The answer lies in the selection of stock birds, and in the nature of the Birmingham Roller. Ernie Stratford, a very successful fancier in the 60's, 70's and 80's, when asked this question made what I believe to be a very telling statement - "You need to remember that the Birmingham Roller is not a team bird by nature". His view was that the roller was an individual performer, and it is a purely human desire to turn it into a team bird performing with its colleagues in unison to earn us points in competition. I would certainly agree to some extent that the very best deep, solid and frequent rollers have very little ability to kit. So when we breed this rather special individual bird, it should not surprise us to find it has difficulty in performing to its best and struggling to keep with the kit. On the other hand there are those birds which are not particularly special, do not roll very deep or frequently and yet still find themselves out of the kit more often than not after a roll. The latter birds are of little use to anyone serious about Birmingham Rollers and getting the most out of a kit performance. A little caution should be taken here though, youngsters when beginning to learn how to roll will often hang above, below or behind the team, often with frantic tumbling and turning away from the team. These birds often make very good rollers when they eventually get their act together. It largely depends on how long you can afford to wait for them, how many of these may find their way back into the stock pen (to create even more of this phenomena), and whether in the process of waiting they invite other birds to leave the kit.

So the answer to the question is very little. Lenny Harris once told me that when a bird starts to leave his kit, after it drops with the team he places it in a cage and lets it see the other birds eat without it. He suggests that the next time it flies with the kit it is likely to stick a bit tighter to them. Personally this has never worked for me, but then I often can't be bothered to try it. Other fanciers suggest that keeping the birds in a smaller kit box or in a darkened kit box will aid kitting, once again I have not found this particularly helpful, but then again I have not persevered with either method. It is important that before you condemn the bird you eliminate the possible transient reasons:

- It is still developing and is learning to roll properly.
- It is ill fed and because it is rolling more than the rest of the team it is tiring.
- The kit has overflown and it is simply bored....after an hour and a half even the best kitters may be seen to leave the kit.
- It has paired to another in the team and is more interested in production than performance.
- It has developed some disorder or trauma which is affecting its ability to kit i.e. a blown eye, a knock which is affecting its flying ability - stretched tendon in the wing muscles, painful joints - arthritis if it's getting older, or a previous injury from rolling within the team.

- It has learned that if it stays out of the team it can control its roll more, has become aversive to rolling and desires to reduce its rolling.

7. I have a lot of birds rolling from the back of the team - can I change this?

These birds are usually the most prevalent rollers in the team, and are tiring quicker than the rest of the team. It generally means that the rest of the team are probably overfed or over fit, and the team is not balanced. You can change this fairly easily:

- Fly the team less often.

- Fly the back of team of rollers less often, and feed them a little extra than the rest of the kit.

- Ensure the team is at the same fitness - either reduce the fitness of the faster flyers or fitten up the birds rolling from the back of the team.

- Balance the team through a programme of feeding and flying - see Q22 on balancing the team.

8. My birds roll sloppy one day then they are okay the next - what can I do?

You are probably flying them too frequently and probably not feeding them enough. If this is not the cause then it will be about their general lack of fitness or poor breeding. The most likely reason is the lack of energy the birds have, and this is due to underfeeding or overflying. Most fanciers still believe that a thin undernourished bird will perform best, and to some extent this is true in terms of frequency, it does not hold out to the same extent for quality of the roll. The Birmingham Roller must not be overweight, but for it to have the energy to roll properly and with quality it must not be underfed. Getting the balance right is the art of feeding Birmingham Rollers, and will turn a good team into an excellent team. Unfortunately it will never turn a mediocre team into a good team! The alternative explanation for this phenomena is that the birds you have acquired are simply not able to roll properly, and they occasionally look good when all the conditions are right - i.e. the wind is from the east, a high pressure day, and the wind is no more than 5mph.

9. My birds roll down when trying to land, is there any cure for this?

When birds are slowing down to alight at the loft is the most dangerous time for high quality birds to roll. The bird is in stall speed, and thus is more likely to want to roll - this is a natural phenomenon. After all if you have watched a kit perform you will have noticed that it is when the kit is slowing or pausing in their flight that most activity happens. So clearly this is the aerodynamic state when the trigger to roll is the greatest. Thus those Birmingham Rollers that have the most roll in them are likely to want to roll when landing. This of course is pretty annoying, as it means that some of the best and most frequent of your rollers can injure themselves or even end up killing themselves. There are 2 main trains of thought on this:

1. You really don't want to perpetuate rollers that do this, if they aren't culled then you will end up with even more doing this, so they should never get into the stock pen, and they should be eliminated from your team as soon as possible.

2. These are valuable birds, frequent performers in your team that are only doing what you have bred them for, breed even more of these and have greater performance in you teams.

If you take the first view, then there is no cure and the birds like this should be either culled or simply enjoyed until they self-destruct or injure themselves beyond repair. (Not the most humane position to take though)

Take the second view and there is one simple solution. Erect a 30 foot landing pole. This will enable those rollers that exhibit this trait to roll when they slow their speed to land and still roll a short way, control their spin and avoid connecting with loft roof and thus avoid injury. I have had many good rollers avoid injury this way, and unfortunately when I did not have the luxury of a landing pole have seen many good rollers come a cropper too many times unnecessarily. The landing pole will not save injury to the 'roll down` or the erratic roller that cannot control its roll at all, but it will save injury to the frequent roller that you can enjoy performing - the roller that is, after all, just doing its job to excess.

10. I have really deep rollers, what can I do to help them in competitions?

Deep rollers will be exciting and interesting to both onlookers and judges, however they will seldom win competitions. Both John Lenihan of Bristol and Peter Stripp of Northampton are competitors that have enthralled me, and frustrated me at the same time. Their birds were absolutely brilliant in their performance of the roll, probably mostly rolling tight and fast for a minimum of 3 seconds (which is very deep indeed). They also had the grace and style giving the spectator that image of a spherical object smoothly spinning straight and true. However the frustration was that they were seldom in the sky at the same place for very long. Hence it was virtually impossible for them to have a winning score because they weren't getting the points for breaks or kitting. Having said this, both Peter and John did from time to time win competitions. How did they do it? Well I believe that they had to select their birds very carefully and feed them well. All deep rollers will need to be very fit to keep with the team for any length of time, they will need to be muscled up but light in order to have the stamina to continue to go back to the team after rolling. Also the rest of the team will need to be flying slowly and steadily for the deep rollers to collect and go back to the team. This will require the competitor to ensure the main team is a little heavier than the deep rollers, is exercised less frequently and fed more millet than the deeper rollers. Millet as you will note is the seed which tends to slow down the wing-beat and if overdone will 'cool` the performance of the team.

11. My birds look 'wingy' when they roll, is there a cure for this?

I think this is generally created in the stock loft. This winginess is developed by the improper performance of the rolling manoeuvre. When a Birmingham Roller rolls it uses a wing-beat which is done so fast that it is imperceptible to the naked eye. The top and bottom of the wing-beat creates either the smoothness or the winginess of the witnessed performance. For example – when a Birmingham Rollers wings touch at the top and bottom of the wing beat it will look smoother than one in which the wings change upward and downward stroke without touching. The wingy motion is created by the bird which does not complete the full stroke at either top or bottom. The less the full stroke is completed the more the bird looks wingy. When a Birmingham Roller is seen from directly below the bird will take on a view of an A and H or an X. The A shape is the most accomplished, the H is a very acceptable spectacle, and the X looks rather gangly and wingy. Sadly I believe that the X shape may in fact be spinning faster than the other 2 shapes, so perhaps putting more energy into its performance, however it is not the spectacle which pleases the eye most, so unfortunately its efforts are wasted.

Now in answer the second part of the question - is there a cure for this, the answer is that I can't say for sure but the answer is probably no! After all I have only been flying and breeding Birmingham Rollers since 1963, so I haven't explored everything to do with them yet! However I have a couple of theories which might help.

It is possible to cross such X type birds with more graceful A or H styled birds and produce reasonable offspring. Such hybrids may actually be more likely to have 'A' style characteristics. How one keeps this quality in the following generations is a consideration for the expert breeder to contemplate further and pay particular attention to the selection of breeders in the 2nd and 3rd generations. I don't yet know of no breeder who has successfully done this cross and continued to produce high quality Birmingham Rollers consistently, although I know many who have had limited success in the first generation of crosses.

Secondly, I offer another aspect for the fancier to consider: whether it is the actual physiological makeup of the bird and its genetic musculature, or the development of the muscles in the rearing and the physical state of the bird in the day to day exercise and training regime of the individual fancier. The second hypothesis offers a little more optimism, as it could mean that a change in rearing and feeding practices may be an influence of winginess.

I have always stated that rearing youngsters and feeding youngsters is important if you want to create good old birds as yearlings and older. So as I have very few (if any) wingy birds, we have to wonder whether it is that I have the right stock birds, or whether is it that I rear and feed my youngsters (and old birds for that matter) differently from some other fanciers. I have always ignored the '1 tin for 20' bird rule, and have often been accused of feeding too heavy. I have noticed that when I have tried to cut down on the feed to increase performance that the quality of the roll has fallen. So perhaps there is not a direct link to feeding and winginess, but it does seem that it is worth further speculation and experiment…..what do you think?

12. My birds roll great for about 5 minutes then just go flat - what can I do?

This is about feeding properly and balancing the team. When a kit has been deprived of its freedom for a period of time - say 4 or 5 days or so, it will be full of roll for the first few minutes (if it has any roll in it), however if the team is not fit and balanced then this burst of energy will quickly run out. The team has to be fit and full of energy to perform for the full 20 minutes and beyond. The half-starved kits which prevail in the UK are such that they will perform very frequently for a short period of time then go flat as their energy runs out. This is because the feeding has been reduced to increase the performance, but no reserves of energy are held by the bird for sustained performance. See the answers to Q1 and Q24 on how to do better in competition and feeding a team to see the solution to this problem, or examine how a runner prepares for a race to get the right idea of training a kit of Birmingham Rollers.

13. My birds can all roll, but they don't seem to roll together - what can I do?

The problem here lies either with the feeding regime you have; the selection of the team; the experience of the team; or the relationships within the team. Let's look at each in turn:

- The feeding regime is really important, so ensure that the team is balanced - see the answer to Q 22 for detail on this. Birds that are of different weights and fitness are much less likely to roll together than a team which is balanced.

- The selection of the team is important as to have too many deep rollers in its composition is always going to lead to greater difficulties. The best team I ever flew was a team that had 4 short rollers, 14 medium rollers and 2 deep rollers in it. Some fanciers prefer to have all short rollers or all medium rollers in them, but I would say this is likely to lead to rather boring performances, although it might pay dividends in competition.

- The experience of the team is important too. Having a few older birds in the team will have a steadying effect and help them fly in a slower pattern, and make them more likely to fly in a figure of 8. How long the team have been together will also influence their concert performance. It is amazing how a team put together over a period of time seem to get to know each other, wait for each other and keep together much better than a team cobbled together over recent weeks.

- Finally, the composition of the team from my point of view is about how the birds relate to each other. My preference would always be to fly a team of brothers and sisters, half-brothers and sisters, cousins, half cousins, as opposed to an unrelated team. Closely related teams are easier to balance, easier to feed similarly, and fly with the same speed, wing-beat and circle as each other. All of these factors make for a greater possibility of `concert' performance.

14. What can I do about the deep rollers? And what is the best depth to do well in competitions?

Deep rollers are a joy and can be very frustrating. Fly no more than 2 in any team if you want concert performance. If on the other hand you are not concerned with competition or concert performance then knock yourself out - enjoy them. To get the best out of deep rollers I would suggest you fly them no more than twice a week, resist the temptation to fly them more often. Ensure that they are fed well, but it is essential that they do not get too overweight. If any panting is noted when they land reduce the food ration slightly and fly more regularly until they regain fitness. If for some reason they have got too out of condition feed barley to increase their fly time and slim them down in the process.

The best depth of roll to do well in competition is 1.5 to 2.5 seconds in duration. More than this will mean it takes too long to return to the kit. It will, of course, depend on the judge too, some judges will want to judge only those deeper rollers and some will be tolerant to short rollers. However the range above should satisfy most judges. George Mason has often said that if the bird rolls long enough for the judge to appreciate its speed and style it is deep enough. A 2 second roll is certainly deep enough for most judges to evaluate it correctly. Some extremely good quality rollers can be very `short' in the roll and even appear to be losing no altitude at all, they will however still be spinning for about 2 seconds in duration. I use seconds rather than `feet or yards' because I really believe that it is far too difficult to accurately assess the distance of the roll unless there is a structure behind the bird to give perspective, whereas counting seconds seems to be a more sensible way to compare `depth' factor in the execution of a roll.

15. What can I do when the kit goes flat?

A good team of Birmingham Rollers cannot be expected to perform continuously for long periods. A team in excellent form can be kept on form by the experienced and skilled fancier for about 12 weeks, and following that the team will need resting and re-balancing before they can be expected to perform to a peak again. Some fanciers may be able to exceed this time frame and most will fall greatly short of this potential. Having had the pleasure of consistent weather and few predators while I lived in Portugal, I know that 12 weeks is possible. It was seldom longer for me, and then only when I was not asking too much of them and only flying them twice weekly.

16. How can I stop my kit from flying too long?

Believe it or not, there will be some birds in your team that encourage long flying. Taking these out will reduce the flying time considerably. If you observe them closely you will notice that when the team looks like landing there are always a few birds which fly above the kit and take them around for another circuit. When the birds first come out of the loft there are always a few birds eager to fly up and they seem to lead the birds up higher - these are the birds that are persuading your kit to fly longer. These birds seem to thrive on less food, seem lighter than the rest and get fitter quicker. These birds are the culprits, take them out and the team will be more under your control. Of course identifying these birds is your responsibility and cannot be done without you watching carefully - something most of the unsuccessful fanciers seem to be averse to!

In addition to finding the culprits on a day to day basis, there is one other factor:

The birds are too light. Believe it or not, birds which are very fit and light in body weight with insufficient muscle tone, will fly longer and higher that those in the 'right' condition. So feeding a little heavier, perhaps even increasing the protein balance of the feed will reduce the flying time, not increase it. Recently I had to fly 3 teams back to back in the All England Roller Club fly and was allowed only 30 minutes between each kit once the time had elapsed, and only 20 minutes if I decided to scratch the second kit. This meant that the teams that had been flying normally for about 2 hours each, had to be reduced considerably. Common wisdom would dictate that the food should be reduced to reduce the flying time, but I knew that that would not help. Dave Moseley, who perhaps has the best knowledge of anyone I know about feeding, suggested this:

- Feed 50% high protein mix (breed and wean or similar) and 50% wheat for 4 days.
- Reduce the ratio over the next days leading up to the competition until they are back to 100% wheat 2 days before the competition. An example of this for a competition in 12 days' time would be as follows:

   Days 1-4   50/50 high protein and wheat
   Day 5   41/69
   Day 6   34/66
   Day 7   25/75
   Day 8   16/84
   Day 9   7/93
   Day 10   Pure wheat
   Day 11   Pure wheat
   Fly

- Also while on this programme it is important not to overfly the birds in order to allow them to gain some weight and reduce their fitness. I held them up whilst on the 4 days 50/50, and flew them on days 5 and 9 only.

The result of this programme was that on day 5 their time had reduced to 1 hour and 20 minutes, by day 9 they flew only 1 hour, and on fly day they dropped 20 seconds before I would have had to have scratched the next kit! There is remarkably very little change in performance on this regime, and indeed much less than one might imagine. If anything there was a slight increase in quality and only slight reduction in frequency. It does, however, demonstrate that reducing fitness, and increasing weight will reduce fly time as opposed to the common wisdom of reducing the food, which in my experience does not work at all.

The fancier should also note that the amount of food given is also a factor on this sort of regime, and it should be in line with the 'normal` amount you usually give to achieve the result you want……thus you are, in this programme, doing nothing more than reducing the time, not making any significant change in their normal performance.

17. How can I avoid flyaways?

Gordon Forbes had the answer to this one - `don't let them out'! Gordon was a specialist at this, he had more flyaways than anyone I ever knew with rollers. He came to the conclusion that it was a combination of pinning out - see Q 20, atmospherics, and having the birds too thin and super fit. It is true that birds that are thin and yet not well muscled may perform more frequently (at the beginning of the fly), but lack the energy and muscle power to get down if the atmospherics are unfavourable. My son in law who is an expert para-glider pilot tells me horrific tales of flyers that found thermals which took them up to altitudes where they died of oxygen starvation or froze to death. This illustrates how easy it may be for a bird or a kit with good lift but without good muscle to be sucked up into a thermal and be unable to fight its way back down. Another good reason for having birds which are not underfed and under muscled.

18. What can I feed for slower flying?

Small amounts of plain millet as a supplement (1 tablespoon between 20 birds) added to the normal diet will slow the wing beat down in proper Birmingham Rollers. However an excess of protein in the diet will increase the wing beat. Flying Birmingham Rollers in excessive wind (wind above 16mph) will encourage fast flying and reduce performance. Bear in mind that millet, whilst slowing the wing beat, will also slightly reduce the amount of breaks the kit will perform. You may notice that they will fly through the `hover' point more than usual - this is known as `dodging it', and is often a result of an overdose of millet.

19. How do I stop the kit splitting?

A balanced kit will be less likely to split, however check the weight of the birds and remove the thin ones and the super fit ones. They are the ones most likely to be causing the problems. Even removing 2 or 3 of these birds will make a significant difference. The cause of splitting is usually that the super fit ones and the thin ones are resisting the roll better than the others, they then lift away from the performers and create the split kit. Go back to basics and balance the kit again (see Q 22), ensuring that the thin ones gain more weight and muscle, and remove the super fit ones and rest them until they acquire more weight and less condition.

20. How can I stop the kit pinning out?

Pinning out is usually caused by the birds becoming too light and super fit. Keep the birds a little heavier and this will be less of a problem. To put weight on to a Birmingham Roller quickly I would suggest you use pinhead oatmeal as a supplement to their normal diet. Barley in small measures will also help - but beware: this will lengthen the flight time and may result in a change in flight pattern to increased raking. Also avoid the use of tonics, mineral grit, and supplements in the water. Use some common sense and avoid flying kits on high barometric days with bright sunshine and no cloud. Those warm, sunny days with rather lovely blue skies may be great for the garden, but are not necessarily great for watching fit Birmingham Rollers.

21. What can I feed for longer flying?

The easy answer to this is to feed a proper measure of wheat, and ensure the birds are not overweight or undernourished. If all of the above is already the case, then ensure that they are not being flown too frequently, if this is still not the cause then feed ½ wheat and ½ barley for about a week and their time will increase substantially.

22. How do I balance a kit?

Dave Moseley should be credited here, for it was he who first made me see the importance of a balanced kit. As well as having the best rollers in the team it is important to know that they are all of the same fitness, weight, and energy level to compete in unison. Contrary to popular wisdom having half-starved Birmingham Rollers is not the best way to win competitions! This is the process of balancing a team:

- Fly the team and wait until all the birds have returned and give them some time to settle down before feeding.

- Feed the team together and the same food - undoubtedly my preference would be good quality wheat.

- Ensure the birds continue to eat until they have eaten as much as they can. This is about stretching their crops, so when they stop eating remove the food and check how much they have eaten.

- At the next feed ensure that they eat as much or a little more.

- Fly the team during this process so that they have the chance to turn any excess food into muscle rather than turning to fat with inactivity.

- Continue this process of `overfeeding' until the birds have acquired a `muscly' feel to them. If you have flown them regularly through this process then their flying time will have increased with their extra muscle, and yet their weight will not have increased much if at all.

- Very gradually reduce the quantity of food given after each fly until the fly time reduces to 1 to 1½ hours.
- Your kit should now be balanced.

23. What can I feed for lower flying?

Give a supplement of pinhead oatmeal to their normal ration. This will substantially increase their weight and reduce their ability to fly too high. Be aware that atmospherics will have a significant effect on the altitude the birds fly at, so it may be necessary to check the barometer and note which reading synchronises with the high flying. If you find a correlation, you may be wiser to restrict kit flying in these conditions. See Q20 Pinning out and Q17 Flyaways.

24. What are the secrets of getting the kit ready for a competition fly?

There are no real secrets, and yet there are some simple rules to follow:

- The birds have to be in good condition and able to fly at least 1 hour without individual birds dropping out. Any training or feeding programme relies on this being the baseline.

- If you are intent on reducing the feed to increase frequency of performance then do this up to the few days before the fly, not directly before the fly. The birds should be on a full ration of food at least 3 days before the fly.

- Balance the team (see Q 22) to avoid splitting.

- Ensure the birds are not too light in condition – this will avoid pinning out.

- Don't overfly the birds before the competition. Most old birds need no more than 3 outings a week. Youngsters, once developed, need only to fly every other day to maintain fitness.

- Restrict flying for 3 days before the fly. This is a general rule, however, your particular practice may be slightly different to this. Some fanciers report that restricting flying for longer before the competition and then flying the day immediately before the competition increases performance.

- Take out any birds that you are unsure of, it is better to fly a reduced number in the kit than curse yourself later for failing to remove a bird you 'knew' was suspect.

- If possible have more than one team in training (on the same feeding and flying regime). This allows for the transfer of birds from one team to another should accidents or issues of fitness or health arise.

- Leave well alone! If the birds are doing okay, resist the temptation to 'just tinker a little' at the last moment. More competitions have been lost by last minute 'souping up' than by leaving well alone. I know fanciers who have tried laxatives, antibiotics, anti-depressants, various stimulants, and even paramyxo virus vaccination to try to increase performance – all with poor results. Any tricks or techniques which you think will work should be tested a long time before the competition – not the night before! The saddest story I heard was the tippler fancier who gave his 'best 3 bird kit' pro-plus tablets before a competition, only to find them all dead at release time.

- I find that a bath 3 or 4 days before the competition can help the final performance.

- Ensure that the kit does not have anything to eat or drink for at least 4 hours before the fly.

25. What is the judge looking for?

What a great question! Not quite so easy to answer. At one time I would have been very clear that the judge was looking for frequent high quality Birmingham Rollers, with breaks of over half the team in a consistent fashion. Now I'm not so sure! The tendency I am noticing recently is an emphasis on frequency, so the small frequent breaks with short rolling birds will do better in competition under this style of judging. I do not, however, approve of this departure from what I consider to be good judging. What the judge should be looking for is a question I feel more qualified to answer.

What the judge should be looking for is:

- High quality Birmingham Rollers that roll together.
- Frequency of performance of about 1 roll a minute or more.
- Duration of roll (depth) of between 1½ and 2½ seconds.
- Graceful style of roll giving the impression of fast spinning balls.
- Graceful wing beat, like butterflies in the sky, flying in a figure of eight.
- Birds that stay close together in the kit and after rolling return to the kit without delay.
- Courteous and friendly competitors who listen to the judge's view and accept his or her judgement.

26. How many birds do I need to do well in competition?

Winning competitions is often down to a combination of good birds, the right judge and fortune with the weather. To do consistently well in competition it is important to have some extra birds to support your main kit. Many countries in the world take a more sensible view of competition and have 'any age' competitions. To do well in any age competition you really only need 2 full teams - probably about 42 birds. This allows for them both to be trained simultaneously with the same feeding and flying regime in order to allow the best swaps between kits when injury, sickness, or accident occurs. However, in the UK if you intend to fly in all of the competitions available you will need many more than this. To be certain of doing the best you can you will need 2 teams of the type needed for the competition. That is if you are intending to fly in a 'yearling' competition to be certain of your best result you will need 42 yearlings. If you intend to fly in 'yearling' and 'old bird' flies then 42 yearlings would still be enough, as yearlings can duplicate as old birds. Similarly if you intend to fly young bird kits a minimum of 30 to 40 would be increase your chance of success. This is because the programme for competitions is pretty hectic, and the more birds you have the more able you are to rest teams in between fly dates. Bob Brown who was perhaps our greatest competitor in the UK, always advocated having as many kits as possible – especially for young bird competitions. Youngsters come on to the roll and go off the roll very quickly, and so to have an abundance of swaps between kits is a very useful resource. Watching which birds are active and which birds occupy the front of the kit and being able to swap these into the competition kit will always give you an edge on other competitors – providing you have a suitable quality of stock. Having said that, if you are a serious and dedicated fancier who takes care in the stock selection, knows what you are doing in terms of feeding, training, resting, and balancing a kit, then a smaller number of birds will give you some success too! Without wanting to boast, but wanting to give a clear example, I have achieved success with a very small stock. When I won the National Young Championship in 1982 and again in 1989 the birds I flew in competition were selected from the first and only 30 birds I bred that year.

27. I have some great rollers but a few leave the kit and fly above them, is there any cure for this?

These birds are generally a nuisance, fanciers hold on to them because they are usually good rollers which don't always do this. They are birds which tend to be quick eaters, want to fly longer than the rest of the kit, and get fitter quicker than others when they have been held up for a while. There is, as far as I know, no cure for this, they will do this from time to time and usually when it is the most inconvenient (sod's law). Removing them from the kit when they can be identified will eventually leave you with a better kit, one that is less likely to let you down when the chips are down, and is more consistently enjoyable.

Flying for pleasure

28. How can I get my locked up rollers flying fit again?

Bringing birds out of the stock loft, bringing them out from being locked up over winter, out of retirement, or from another fancier who hasn't flown them for a while requires patience and care if disaster is to be avoided. The Birmingham Roller will be overweight and unfit, it may even have grown a slightly longer tail, its wing muscles may have become atrophied (shrunken) and it may even have lost volition to fly much at all. The first step is to reduce its weight. This can be done by reducing its ration of wheat, restoring it to a wheat only diet, or changing its diet to Barley, Wheat and Dari (milo) mix. In some areas a `depurative' mix is available commercially and this may be a short cut for some fanciers. As well as changing diet the bird must be retrained carefully, much as if it was a youngster being settled for the first time. Bear in mind that no matter how unfit the bird is, if it is a proper Birmingham Roller it will roll at its first opportunity of free flight, which may easily result in an early demise for the unfortunate creature. Therefore it is imperative to ensure its weight is reduced before free flight is permitted. Once its weight has been reduced sufficiently, if the bird is worthy of its label Birmingham Roller, it should have enough control to land safely.

The safest process is therefore thus:

- Place the bird or birds in a settling cage and allow it to see the loft and other birds flying and returning.

- Once some weight has been removed allow the bird freedom from the cage to the loft only.

- The bird itself will only want to fly short distances from ground to loft top, cage to floor, loft to cage etc.
- Allow this for as many days as necessary until its muscles seem to be freer and its energy appears to have returned.
- At some point it may decide to take a circuit of the loft area and return: this is a sign that its motivation to fly has returned you are in business.
- At this point if you have a kit trained the bird can be hand launched into the team when you estimate that they are a few minutes from completing their fly.
- If no errors are made in this flight it is then possible to release the bird or birds into the team for increasing lengths of time until the bird or birds return to a normal fitness and can be flown with the team from the beginning of their release.

29. How can I avoid roll downs?

I really wish I could say that this is completely possible, however, I think the occasional roll down is unavoidable if you want to breed 1st class rollers. While I lived in Portugal for 8 years I only had 3 roll downs and 2 of those were from the same pair and in the same year, so it is possible to minimise these sad mistakes. You are able to do this only by being extremely careful in your breeding programme, and by breeding for birds which are later in developing rather than earlier. In Portugal I had the benefit of being in no rush for the birds to come on, I suffered very little loss from predators (falcons were seasonal, cats and other hostile animals were not prevalent), and losses from human deviants (thieves, scroungers, non-returning borrowers, and racist fire setters) were non-existent. I was therefore able to have a very small selective stock that had only to please me. I had to breed very few, as my losses were almost none, and I felt no need to breed extra birds to supply local fanciers, of which there were very few indeed. In fact my biggest problem was finding good homes for the good rollers I was breeding, as my loft could only contain 4 teams and had 2 breeding compartments.

So although it is possible to have few losses, I think that those of us interested in breeding increased numbers of good fast high quality Birmingham Rollers are going to suffer the occasional roll down. To avoid them completely is unlikely, to reduce their number you need to:

- Select stock very carefully.

- Select late developers rather than early developers.
- Ensure youngsters are exercised regularly and their nutrition is catered for properly.
- Keep accurate records to screen out possible mismatched pairings – erratic overly frequent and deep rollers aren't paired together.
- Avoid very close pairings such as father/daughter, mother/son, brother/sister.

30. How many pairs should I breed from if I'm not interested in competition?

Only as many pairs as you are sure will breed you good quality Birmingham Rollers, and a couple of pairs which are experimental and might breed you better ones. The simple rule of thumb is that each year you must breed from your best rollers, and as all your stock is going to be getting older, a few of your younger birds (yearlings and 2 year olds) to test out. The new introductions to the stock loft have to be at least as good and able to reproduce good rollers as your ageing stock. To have a reasonable kit or two of rollers you will need to breed 20 youngsters a year to replace the losses from peregrine falcons, sparrow hawks and other predators. If your stock is sound this can easily be achieved through having 4 pairs of established rollers and 2 pairs of new introducers on trial.

31. What's the best way to learn more about the Birmingham Roller?

George Mason once said that you need at least 2 lifetimes to fully understand the Birmingham Roller – unfortunately we don't come by another lifetime very easily! So in my view by far the best way to learn is to listen, watch and say little. Listen to the fanciers who have been around for a while, those that are doing well in competition, and those who show the quality of their rollers in the air rather than their stories about how good they were yesterday. Observe as many kits of rollers as you can, you can learn even from the poor quality ones. Watch how the successful experienced fanciers feed their birds, how do they handle them, and in what conditions do they keep them in. There will come a time when you can ask a very sensible question which will show how keen you are to learn, how interested you are, and how much respect you have for those mentors in the hobby. Save those questions for the puzzles that you can't have answered through observation and careful listening.

32. What is the best feeding regime for youngsters, and does it differ in old birds?

I have remarked on many occasions that the youngsters you breed today should be the stock birds you will need tomorrow. So bearing this in mind, it is extremely important that you rear, train, fly and feed them properly. Youngsters are in the process of growing and maturing, and need a good constant diet to fulfil that task. Good quality wheat is sufficient to nourish a youngster once weaned, and it needs little else but clean water, grit and minerals to grow into a good breeding specimen. However, the protein value of wheat may vary considerably, and although most will have about 11%, that is just the minimum requirement for robust growth and health. Thus, especially in the first few months of its life, it is prudent to augment the youngster's diet with both a few high protein foods and with fat soluble vitamins which are not contained in wheat alone. Once a week, or perhaps every 10 days, I like to give my youngsters some mixed seed (millet, canary seed, linseed) and if I am unsure of the protein value of the wheat I might feed a few peas, tares or beans.

Once the youngster is beginning to perform (usually between 5 and 9 months) it is safe to change its feed to wheat alone providing this is given in sufficient quantity to ensure full nutrition.

Yearlings and old birds will also benefit from supplements from time to time. When under pressure in competition periods, a little seed, vitamin supplements in the water, and mineral grit will ensure peak fitness. Ensure these are given after the fly though, as it is not a good idea to give these in the few days preceding a competition fly.

During the winter maize is helpful and during the moult mixed seed for vitamin complement, and peas and beans are useful to ensure good growth of strong healthy feather.

33. My rollers are great as youngsters but turn to water as yearlings - is there a cure?

Turning to water is a term usually to mean that the fast tight spectacular spinning ability of the young Birmingham Roller has deteriorated to loose, often uncontrolled deep rolling. Often described as rolling like a rag or loose doughnut, spiralling towards the ground and often unable to stop with grace or style. Unfortunately I know of no cure for this, the problem lies in the stock loft usually, or in the poor rearing process. Before condemning the bird though, ensure that it is not suffering any illness or debility through poor nutrition, housing or other forms of neglect.

34. How often should I fly my kits?

Rollers benefit from rest and exercise in the correct proportions. Old birds should not be flown more than 3 times a week. However, you need to understand that keeping them in more than that will have a detrimental effect on their ability to kit tightly. The careful owner of a good kit of old birds will monitor the fitness of the individuals and ensure that they get the exercise they need to perform optimally. From time to time it may pay to rest an old bird team for between 3 and 5 weeks to restore the birds to their best quality. Take caution here though, as care is needed as they may need to be dieted before flying them again after this period of confinement to avoid calamity.

Youngsters generally need flying daily, although this needs also to depend on the weather. In Portugal where the weather was consistent, the youngsters thrived on daily flying and developed good habits, strong musculature, and good health. Flying in the UK in strong winds is to be avoided when possible, as this encourages fast flying and poor habits. Once the youngster begins to roll in good style and frequently it may need more rest, and individuals can benefit from being transferred to the yearling or old bird kit if they appear to be suffering from overflying. The symptom noticed in this regard is the bird losing weight, landing early, or persistent short rolling/tumbling at the back of the kit.

Yearlings benefit from slightly more flying, and I generally fly them every other day, and if the frequency of performance begins to decline, I may rest them an extra day. Some strains of birds benefit from more frequent flying, some less - experiment to ensure that you have the optimum for your strain of rollers. Keeping good records of what you have fed, how often they have flown, the state of the weather, and the level of performance is a good habit to get into in order to `master' your management of your strain of roller.

35. What effects will the weather have on flying my birds?

Once the roller has developed properly, the weather will have very little effect on the bird itself. It will, of course, damage the spectacle which the hobby is all about. It would seem foolish to me to fly birds in adverse weather, then go inside to watch TV, and let's face it who wants to watch rollers in strong winds, pouring rain, and freezing cold. OK I guess the obsessed like me do! However certain weather conditions need to be avoided:

- Snow is a no-no! I have in the past cleared the top of the loft off and flown the kit, and have managed so far to get them back, but there have been scary moments. Should the snow begin to fall while the birds are flying they will invariably become confused and will drop wherever they can.

- High winds simply take the rollers into the bad habits of flying too fast, low flying and raking.

- Light rain will actually soak a bird's feathers quicker than heavy rain, and a roller in good condition can survive and fly through a rainstorm better than drizzly damp conditions.

- Fog and mist, if you want to lose a kit of rollers flying them in this is probably the best method.

- High barometric pressure days (hot sunny days with no cloud and blue sky) – a really good way to help your birds pin out and see how far away they can be reported or, as in many cases, never be heard of again.

- Down winds – even in light breeze it is inadvisable to fly kits when the smoke from chimneys is blowing down toward the ground. Rollers will struggle to lift in these conditions and even safe and good rollers can be lost or injured through silly bumps and crashes.

The best days though are the:

- Crisp cold windless days with a frost and blue skies – yes yes yes, give me these rollers to fly every day!
- Rainless but damp days with a slight easterly breeze.
- Sunny warm days with puffy white clouds (cumulous clouds) and slight lifting breeze.

36. How can I help the birds in the moult?

Confine your flying teams until the primaries are moulted. Give your rollers daily baths, feed new wheat if available, and a rich protein diet with supplements of vitamins, and minerals.

37. How long can I expect a kit to perform at their peak?

A careful fancier flying one team with regular rests and flying the teams no more than advocated in Q34 can expect the kits to perform reasonably well for about 12-16 weeks. In order to have them at their peak it is not possible to keep them 'on the boil' for more than about 6 weeks and that will take an experienced and quite expert fancier. This is why the fancier who has 2 teams in competition training will tend to win out over his or her competitors that try to use just one team. There is a little variance in this, as yearlings will tend to last a bit longer, old birds tend to run out of steam a bit earlier, and youngsters are a bit unpredictable but generally once they start performing will continue until they begin to moult their end flights. I do say they are unpredictable as they are also quite capable of going through a 'quiet stage' in their development, and also predicting a young bird kit's performance will assume that they are all about the same age, if this is not the case then predicting performance of mixed aged youngsters is as reliable as throwing a dice. Ideally to have continuous performance I believe it is necessary to have a mixture of all ages in the kit, and being able to swap and change the team to freshen it up is the ideal. It is my view that we would all see much better performances in the UK competitions if we competed with mixed age teams than with the currently inflexible age related competitions. Having said that please do not construe from this that I advocate going back to an 'any age' fly when most teams comprised of all youngsters, but I do believe a few youngsters in a 'best team' competition can create some frequency that balances nicely with consistent quality from the yearling and old birds within it.

Breeding for consistent high quality rollers

38. What's the difference between 'in breeding', 'line breeding' and 'out crossing'?

Inbreeding is the pairing of relatives together, the term refers to any pairing of any relative -even distant relatives such as 2nd cousin to a cousin twice removed. It implies no system, or thoughtful process. Generally fanciers who randomly pair relatives together are more likely to perpetuate the mistaken beliefs (myths) that inbreeding is a bad thing and causes mutations, weakness, deformities and such like.

Outcrossing is the pairing together of birds that have no blood relationships at all. Even distant relatives will be selected out from pairings to avoid mating relatives. In most species of animals that have been bred selectively over a period of years, it would be almost impossible to find matings that fulfil this strategy. In Birmingham Rollers, not only is this practice extremely difficult to achieve, but as one proceeds to the 2nd, 3rd, and 4th generation it becomes impossible to perpetuate without continuing to acquire foreign stock from far and wide. It would seem unproductive to seek out unrelated stock, which will probably be inferior, to the birds you have already available, known to you, and likely to lead to more consistent progeny.

Line Breeding is the purposeful pairing of relatives, to a system and process designed to accentuate the qualities of the animal, and reduce the deficits or faults in the progeny. Fanciers breeding 'down a line' will have in their mind the outcome 'type' that they wish to produce, and will favour 'pairing back' to a particular cock or hen bird which offers the closest to the `perfect' specimen that they hope to recreate. This may mean that granddaughters and grandsons, great granddaughters and great grandsons will be paired back to their respective ancestors (grandfathers, grandmothers, great uncles, great aunts etc.) to a particular plan laid out either formally, or simply in the mind of the fancier controlling the process. From time to time within this process the fancier will probably try to increase a particular quality or reduce a deficit by closer matings such as ½brother/½sister pairings, uncles to nieces, aunts to nephews, in an attempt to add flavour to the mix. The careful `line breeding' fancier takes care in doing this, and will keep accurate and detailed records to ensure that `random' factors do not enter the `line' and produce ill effects instead of the intended improvements. In Q39 you may note that using a bull cock method will accelerate this process, but brings with it some disadvantages which must be considered before embarking on the practice.

39. What are the advantages and disadvantages of the Bull cock method?

The advantages are that you will have a close knit family; the teams that you fly will be closely related. This means that they are likely to fly in a similar pattern, have similar wing-beat, and have the same metabolic rate so feeding becomes easier to balance kits. The quality of the bull cock will feed into all the qualities of the hens to which it is put. If you continue to use this system it will not be long before the qualities of the bull cock will be seen in all the Birmingham Rollers you fly. You will eventually produce teams of birds which are uniform to the point of being identical - you will achieve a very high level of consistency.

The disadvantages are that you are closing the gene pool, and so your original choice of bull cock is highly significant. If your bull cock has some enduring fault, this fault will be transferred to each and every bird in your stock very quickly. Additionally it will restrict the type of Birmingham Roller that you fly: the system relies on wanting to produce uniformity, and this may result finally in your Birmingham Rollers being very similar in roll style, colour, quality and frequency. Finally it may be difficult to find sufficient hens that are complimentary to the cock bird. I spend a long time trying to find the most complimentary hen to its best counterpoint cock, so finding six or so hens for one cock may be less easy than you may think.

40. Inbreeding: is it really necessary? Is outcrossing better than line breeding?

Yes, inbreeding is definitely necessary if you wish to perpetuate high quality rollers that perform in unison without indiscriminate faults throwing up randomly. And is outcrossing better? Well no - definitely not! Outcrossing will be the ruin of most families of Birmingham Rollers. The outcross will bring with it lots of unknown gene patterns, any of which may stop the perfected rolling behaviour being created. From time to time bringing fresh genes into the gene pool will be necessary to create more vigour and resistance to illness and potential genetic abnormalities. However this needs to be done very carefully using a bird from a line bred family of top performing Birmingham Rollers. The most sensible way to do this is to introduce a line of hens from another family because hens' progeny can be accurately traced and secured. To be even more certain of not introducing poor genes into your perfected gene pool I would suggest you then only use hens produced from the outcross for 3 generations to ensure a good cross was made.

41. How do I know the stock is the best I can get? How do I find the best stock?

The short answer to this one is to breed it yourself. For very logical reasons the best stock will not be available to the newcomer, or even a long standing fancier who has not been doing well already in competition. The logical reason for this is twofold: the successful breeder of Birmingham Rollers is just that because they have kept only the very best stock for themselves. This might seem selfish, but if you consider this for a moment it does make perfect sense. If the expert breeder sold or gave away their best stock they would not be in a position to sell or give away the good stock (probably better than most others) that they consider just less than best. More or more selfish reasons also apply, for it is true that some expert Birmingham Roller breeders do not want to share stock with others and then have to compete against their own efforts. Fortunately there are fanciers like Bill O'Callaghan, and Dave Moseley who are more interested in the hobby than winning competitions - it is people like this who are my heroes in the hobby, for it is those people who lend out, sell or give stock away. Some other fanciers would not be concerned about the future of the hobby but would hike up the price and still sell substandard stock. These latter fanciers are the ones to avoid like the plague!

However, there are some simple rules to follow when trying to acquire better stock:

- Know what it is you need - a good hen, a good cock, one pair, or two pairs.

- Have a standard of roller in your head......short fast, graceful deep, frequent, high quality, or even so much roll it can hardly fly!

- Know the fancier - it needs to be someone you know about and trust, be aware of what their sort of Birmingham Roller has to offer you, know or at least chat to them about the faults they find frustrating with them, the things they have to be careful about before stocking, any prevalent flying faults they have to screen for and so on. Believe me, no-one has perfect rollers straight from the stock pen to the air!

- Decide the minimum amount of stock that will be useful to you, and be prepared to pay as much as is necessary for it or them. Obviously if you have made a friend of this fancier it is preferable to borrow stock. This is because the promise of it returning will ensure that you will take home a much better bird than if you have paid for it and thus permanently deprived its owner of its potential.

- Make the most of what you have acquired. Use feeders to 'milk' as many youngsters as possible from the acquisitions. At this point it is necessary to breed quantity in order to see all the potential that you have acquired. Fly as many as possible to see any faults and their best qualities so you can select the best for pairing back into the family to strengthen your stock.

42. If you were starting again how would you go about breeding a family of rollers?

I did this about 12 years ago, when I had decided that my own strain of rollers had deteriorated beyond hope of resurrection. With losses from thefts, flyaways, moving house, loaning and gifting too many good birds, the quality of the stock was too poor to bring it back. It would have been possible, of course, to start again from what I had, but I determined that I would have to spend too many years with little to watch in the process and couldn't see the point of re-inventing the wheel. I had spent a couple years previously judging and travelling on the All England, and National flies, so I was in a good position to know where the good birds were located. After looking at the fanciers winning competitions, and the style of rollers they kept, I kept coming back to Dave Moseley's and Bill O'Callaghans' family of birds. They attracted me because they had some of the properties I admired in the Birmingham Roller. They are first class in the roll, round graceful and fast, kit well, work well and are generally intelligent and easy to manage - not wild, wilful, or difficult to settle. They were not in my opinion free from faults, as I had noted the Harris family had a tendency to lift, to sometimes split, and when confined for some days their kitting ability decreased. I was however, quite happy to either accept these disadvantages and/or work on them further. I knew also that Dave had already reduced their tendency to `pin out' through selective breeding, and Bill had consistently bred for a closer kitting family.

So these are the essential first steps:

Know what's about, know what you like, and see what's available.

From there, for me, the latter was very easy - both Bill and Dave are personal friends and were delighted to help. That year Dave gave me a total of 78 birds to fly and try out, and Bill gave me several excellent stock birds and an invitation to borrow or have anything I needed. There were many other offers from good friends too many to mention, of which I was very grateful, but decided to gracefully decline. One generous gift of 10 youngsters was from Les Bezance and Dean Forster. These youngsters turned out to be excellent rollers, probably faster in the roll than the Harris birds, but on balance did not have the gracefulness that I preferred in Bill and Dave's family. Another helping hand was offered from George Mason, from whom I had had various gifts over the years, which for some reason or another had not worked out for me. So really think it is very important to note that it is not disrespectful to prefer one type of roller over another, and certainly when deciding on what you want to start a family with, you need to be clear about what it is you like. Some fanciers will want fast frequent short rolling birds, others more graceful (but perhaps marginally slower), or deeper in roll. Some fanciers will have preferences for colour, eye colour, clean or feather legged, crested or non-crested, high fliers, low fliers etc. Your preferences must be foremost in your mind before you even contemplate beginning to propagate a family of birds.

To answer the question more precisely, to produce a family from good basic stock is easy. I observed them closely, I picked only the ones I really liked to breed from, I test bred from a lot of speculative pairs, and culled anything that demonstrated an unacceptable fault. Read on to Q 50 for a more specific answer to the question 'how do you do it'.

43. What are the qualities necessary in a flying bird before stocking it?

A Birmingham Roller must demonstrate its ability to roll properly according to the criteria set out in Q4. Clearly it may be difficult to fulfil all 10 criteria at once, however the closer you can get to all 10 the better chance of success in the stock pen. The bird must also be a strong healthy bird, with good feather and a good constitution. If you simply take out the best rollers in the team, you may find that the specimen you have chosen is not a good choice to further you breeding plan. Especially when line breeding, and even worse when indiscriminately `in breeding' (see Q 38) it is important to select healthy vigorous well feathered birds for the stock pen. Any deficiency in the actual physiology of the bird is likely to be a weakness perpetuated in the youngsters produced.

44. Should I go for quality or frequency?

I really should study my answer to this question, as it would seem that I need some advice on this issue. I have consistently gone for quality, when presented with such a great gift from Dave Moseley of some wonderfully frequent and high quality birds, I have over 6 generations reduced them to rather `stiff' high quality birds. I really need to remember when selecting stock to breed from the importance of frequency. Of course frequency is important, for without it the enjoyment of watching a kit perform is drastically reduced. The successful fancier must find a worthy balance between the two. While in Portugal and having no competitions to concern me, it was easy to mix youngsters, yearlings and old birds into a mixed team that would create the energy of frequency and the excitement of quality that I desired. Since returning to the UK I have to think in terms of exclusive teams of youngsters, yearlings and old birds, so I now have to take frequency into a greater account. I have something extra to say about this in Q 49. So my long winded answer to this question is always go for quality first and frequency second, BUT remember the balance is very very important if you are to be entertained daily and are to have success in the competitions.

45. How can I test out my breeding without breeding lots of birds?

Simple, be generous. Lend, or give only your best potential breeders out, with the agreement that the lucky recipient gives you feedback on how well the youngsters produced are performing. Be even more generous, breed a kit or two for your trusted fanciers and let them fly them for you, again with the agreement that you need to know precisely how they turn out. If you want to be less generous you could contract to have the pick of the youngsters after the loan or gift, or ask them to breed a couple of youngsters of your choice from the best ones bred.

46. What are the faults in flying birds that will show up in the breeding loft?

A very common question, but before I list the ones that I am sure of, just be aware that some of the faults listed may have been caused by poor training, or poor rearing. If you are sure that neither of these causes is possible, the 10 most enduring traits below are in my view in the genes and thus will show up (eventually) in your youngsters produced.

- Poor kitting.
- Loose rolling.
- Winginess in the roll.
- Uncontrollable rolling - dangerous rolling on landing and taking flight.
- Rolling down to the ground from wherever they begin the roll from.
- Tail riding at the end of the roll.
- Fast flipping at the end of the roll.
- Turning away from the kit at the end of the roll.
- Mad tumbling as a youngster before maturing into a proper or reasonable roller.
- Laziness - dropping away, dropping early, reluctance to fly.

Some of these faults may be 'ironed out' by outcrossing a different family into your stock, and certainly I have seen this done successfully for the first generation of produce. However, I am not convinced that these faults are so easily controlled in the following generations, and I believe will require a high degree of expertise to ensure these faults do not re-occur in future generations. I think you will need to be a good geneticist, excellent record keeper, and thorough stockman to avoid the risks of these faults rearing their ugly heads in the future once established in the stock pen.

47. How long should I persevere with the stock I have?

This depends entirely on how attached to your family you are! It took me several years before I was able to abandon my old family, admit defeat, and begin to establish another based on the work of other excellent fanciers. Equally one might be persuaded by a remark George Mason once made to me, which was 'if I see something better than mine, I will change to it, but I will have to be sure it is!' Well I agree with the sentiment entirely, especially the last bit - I believe too many fanciers are a little too eager to abandon their own stock when they see something better, or are told there is something better elsewhere. Currently I notice fanciers changing their family of birds quicker than they may change their minds. Pursuing whatever is the flavour of the month can be costly, not only in terms of the financial implications, but also in terms of time and enthusiasm. It must be disconcerting to keep changing horses in mid race and to achieve no success. To answer the question more definitively, I believe that to seriously do justice to any stock you need to breed from it for at least 4 generations. Bob Brown was once heard to say that no new starter with Birmingham Rollers should expect success for at least 4 years. He based this on the premise that even if you get a good start with good stock, you can only feel truly successful when you have bred, trained and flown the birds you have created yourself. Buying success is of course one way to go, but I know of no-one who has bought a good kit of birds from a successful fancier and been able to continue that success without putting in the work to breed good rollers from the stock purchased. The fancier who supplied the birds did not also sell their expertise, their knowledge of the family, and what to pair to what to maintain the quality of the stock released. You have to learn that for yourself. I know from my own experience that it often takes 4 years to appreciate the value of one bird placed in stock. It is often not the immediate offspring which might impress, but the grandsons and great grandsons that demonstrate just how good that bird was as a breeder. So yes: be impatient to do well, experiment with changing matings as often as you like, but also have realistic expectations. The stock provided (if it was the best you could get) deserves a chance to show just how good it is; it may need a little time and your developing expertise to prove itself worthy of your patience. If you have bred nothing of note in 4 generations then it is unlikely you will, but even if you have only bred one or two outstanding individuals, then the genetics are there for you to take advantage of, do so and get the buzz from creating your own family to be proud of.

48. What's the best method of selection?

I have always believed that selecting from the performance in the air is the very best policy. Observing the bird's performance over a period of time, noting any faults, irregularities, and recording accurately its development will save disappointment later. It is possible of course to select pairing by pedigree, and to some extent if you are line breeding then this will have to be part of the equation. However if the bird fits into your planned breeding, it is a healthy specimen, and it is performing well in the air, then it deserves its place for a trial in the stock loft. Bill Barrett often said that he paired up on the ground not from the air, but he also would not consider a bird for the stock pen unless it could roll to his satisfaction, or was so well bred that he had the greatest of faith in it. By pairing from the ground, he meant he would watch the birds as they picked around on his lawn, having a bath, and generally `playing around'. He would decide on which hen for which cock during this process, so he would take into account things like the vigour of the specimen, the size, shape, colour, eye colour, wildness, alertness, and expression of the birds at play. All these factors could therefore be considered as he mentally paired one to another, and like some instinct would `just know' which to pair to what. Bill Pensom often wrote that he liked to and was able to identify the `best' rollers by handling them in the loft, and by noticing the `expression' in the bird's eye. Now after flying, breeding and competing with rollers for 50 years I think I know what he meant. There is something special and striking about a bird that will `click' with another and produce a good pairing, it is as the French say `je ne sais quoi - 'I know not what', it is a feeling, a sense, and an intuition that seems to come with experience. Perhaps not too helpful in this sort of a volume, but nonetheless when you know what I mean you really will know what I mean!

These days with the increase of raptors and other predators it may be necessary to breed blind from time to time. When forced to do so, remember a few of the following points:

- Use the naturally small specimens where possible, but not because they have been undernourished or poorly reared.

- Use hard colours to soft colours - I remember Ken White once saying this, and at the time I scoffed at the notion. Now I know differently. Using Chequers and selfs to white flighteds, badges, balds, oddsides and silvers will help to maintain frequency and quality in better balance. I acknowledge that this is hardly scientific, but I cannot in conscience not mention what I have observed consistently with my own eyes.

- Some fanciers consider that eye colour is also significant. I cannot say for sure whether this should be a consideration. What I would say is that if I were acquiring stock from a fancier who had produced good results, I would listen to his or her beliefs about any particular factors that may influence success with his strain and let that be a factor in my future selection of pairings.

49. What are the qualities in a roller that make it a good breeder?

Well I have a surprise at the end of this question that not all fanciers will agree with, however, after 50 years I suppose I am entitled to one surprising speculation. But first I'll answer rather philosophically to this one. What makes a really really good breeder is quite a mystery. The qualities are easy, they have been enumerated over and over again throughout this volume, they are the same as what makes a `perfect roller in the air' but please note that a breeder has a few more qualities than just the perfect roll. The old adage `the proof of the pudding is in the eating' is right on the money here. Some of my most prestigious breeders I have produced have been birds which I hadn't really considered to be the best. I have had successes with the ones that look right, have been bred right, and the ones that fit the breeding programme, yet the really really excellent ones are often a surprise. Sometimes they take a while to find, and sometimes even years.

The really good ones are often discovered in retrospect, when you look through your breeding records and find the same ring number occurring over and over again. Bill O'Callaghan this year discovered that his exceptional birds nearly all originate from 2 cocks that he could have used much more had he realised. His `drop dead' cock and his `canker' cock seem to have bred him most of his current glut of excellence. So it is, that often when we search for the qualities of a good breeder we find it not in the air but in the records. A bird that can produce good rollers consistently, that its offspring can produce good rollers, and its offspring's offspring produce good rollers is what makes really good breeder! Okay, so maybe that doesn't help us to recognise and utilise them straight away, so what does it tell us? It tells us that when we find an exceptional bird we should stock it for a season no matter what its age is, even a youngster that we haven't evaluated fully, providing it has been bred right and fits into our plan. We should then take as many offspring as possible from it, and then and only then risk flying it again to further prove its worth. At least that way we will not be completely heartbroken and regretful when or if the Peregrine Falcon takes it out of the sky. If its youngsters begin to show promise, then we should remove it from the air and place it in its proper place in the stock loft until its youngsters prove themselves or not. If the youngsters begin to show signs of excellence, then we should select a few of them, and test out if we really do have a champion breeder to look to. If in the end it is really so exceptional we should pair it to different hens or cocks and see if it still produces. If all this proves positive by the time this bird is 5 years old we have a strain maker, and one which we can exploit for perhaps 5 more intensive years of breeding excellence. Finally to give you a little more to go on I elaborate below the ideal qualities I look for in the birds I place into stock.

- The bird is near perfect in the air.
- It has never let me down in a competition by leaving the kit or landing early.
- It rolls at least every time the kit breaks.
- It flies in the front of the team and often begins the breaks.

- It drops with the kit, never flies after the kit has landed, never drops first.
- It fits into the hand and slides through it like silk.
- It is small rather than large, is proportionate (not lengthy in the tail or deep keeled) and apple bodied.
- It has an intelligent expression and always looks alert.
- It is calm in the kitbox, and handles without struggle and panic.
- It seems to always look well feathered, even when some of the others look a bit down with their head in their shoulders. It seems to avoid getting it wing flights and tail soiled when in conditions that every other bird seems to do so.
- This bird rolls consistently well and frequently when it is overweight, underweight, and perfect weight, when it has been flying for 10 days, 2 months, or 6 months.

And finally the surprise I promised you. Both Bill O'Callaghan and myself agree that from time to time it serves the family well to put something a bit `dodgy' into the mix. By this, and against all the other things I have said, from time to time in order for the family to remain exciting rollers, a bird that is `not quite right' pops up and waves at us to try it in the stock pen. It may be one that breaks the rules by being early to drop, rolls a bit too much, or is erratic in landing or taking off. It has (for some indescribable reason) that `umph' - that quality that just says `if you want special rollers use me!' Of course these do not always work out, you may end up breeding a crop of rolldowns, or even dodgier youngsters, so accurate record keeping is the order of the day here, however sometimes the risk will pay off and take your `line breeding plan' a quantum jump to an even better state.

50. Is there a good proven breeding programme I can use?

Well yes and no. There are many 'line breeding programmes' but you can only get out what you put in, so it largely depends on the stock you start with. Line breeding programmes establish and consolidate the qualities throughout the family of the foundation birds. It is possible to consolidate the qualities of 1 bird or over a period of years establish the qualities of many birds first introduced. This is the rather quick and dirty starting point, which is a very strong 4 generation ½ brother sister cross back programme. I would recommend is as follows:

- Take your best 2 cocks - call them A and C

- Take your best 2 hens - call them b and d

- Pair A to b and C to d and produce as many youngsters as possible (using feeders), and test fly them.

- Swap pairing - A to d and C to b and produce as many youngsters as possible (using feeders), and test fly them.

- You should now have a lot of youngsters flying, of which you need to select whichever you think are the best. All the produce of Axb are ½ brother/sister to all the produce of Axd if these are now paired together you are consolidating the qualities of A

- All the produce of Cxd are ½ brother/sister to all the produce of Cxb if these are now paired together you are consolidating the qualities of C

- All the produce of Axb are ½ brother/sister to all the produce of Cxb if these are now paired together you are consolidating the qualities of b

- All the produce of Axd are ½ brother/sister to all the produce of Cxd if these are now paired together you are consolidating the qualities of d

- Now you have some choices to make, whichever birds please you best from whichever pairing, can be paired together ½ brother to sister to favour one of the foundation parents. Do this with at least 2 combinations to prevent the next stage getting too close, i.e. Axd youngsters paired to Cxd and Axb youngsters paired to Cxb youngsters. This favours Cock A and Hen b. Of course if your results overall have been good you may decide to give all the combinations a try and see what they produce.

- From the produce of the ½ brother/sister pairing, select the best birds to pair across the other combination, and select the best of these to pair back to the original 4 foundation stock.

- And of course during this process you may wish to continue to breed from any pairings that have created good rollers.

Diagrammatically the programme looks like this:

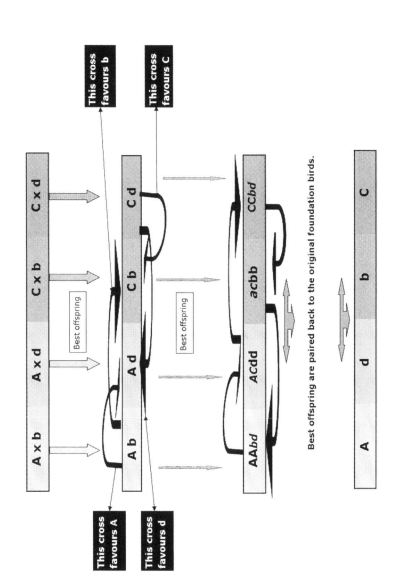

Less experienced fanciers keep asking me questions, what should I do?

50 Questions for the Less Experienced Roller Flier is a beginner's guide to the hobby and answers those questions which are more common among newcomers to the sport. The print version can be purchased at any good bookstore, or online via lulu.com for a 5% discount.

What about the missus?

A Waste of Good Weather by Janice Russell is a light-hearted novel set around the lives of Stan, a Roller Fancier, and his long-suffering wife Evie. From Middlesbrough to Amsterdam, pigeons to diamond-smuggling, this is a story that every fancier's family will want to read and enjoy.

Printed in Great Britain
by Amazon